AI PROMPTS FOR SELF-PUBLISHERS

COLLABORATING WITH YOUR NEW PUBLISHING ASSISTANT

MICHAEL MCCANDLESS

BOOK 5 IN THE PUBLISH IT YOURSELF! SERIES

TOUCHWOOD PRESS, NORFOLK, VIRGINIA

AI Prompts for Self-Publishers: Collaborating with Your New Publishing Assistant

Book 4 in the Publish-It-Yourself Series

Published by Touchwood Press, a division of Touchwood Ventures LLC, Norfolk, Virginia, USA.

Paperback ISBN 13: 978-1-946313-69-0
TBN: P26004P

Manuscript Version 26.4

Contents

HOW TO USE THIS BOOK

This book is an AI tool kit for authors and self-publishers. Use the Quick Reference Guide below to find topics of interest and check out the related prompts. Rinse, repeat.

Writers and self-publishers will want to read our book *Artificially Intelligent* for a more complete treatment of the issues, tools, challenges, and opportunties related to AI and what it means for the art and craft of creative writing and publishing. Get *Artificially Intelligent* on our website, touchwoodpress.com, or on Amazon.

The prompts are designed to be copied, customized, and used immediately in ChatGPT, Claude, Gemini, Copilot, or any other conversational AI tool. They are written in plain language — no technical jargon, no complicated syntax. For more information about choosing the AI platform that will benefit you the most, read our book *Artificially Intelligent*.

Using the Prompts

The bracketed placeholders — like [YOUR GENRE] or [BOOK TOPIC] — are yours to fill in before submitting the prompt. The more specific you are when filling them in, the better your results will be.

A note on iteration: No prompt produces a perfect result on the first try. Think of these as conversation starters, not vending machine buttons. When a response isn't quite right, say so. Tell the AI what you liked, what missed the mark, and what you need instead. The second and third exchanges are almost always better than the first. Wash. Rinse. Repeat.

The Golden Rule of Prompting: Give the AI a role, a context, a task, and a constraint. Many of the prompts below already do this for you. As you grow more confident, you'll start writing your own using the same structure.

Note that there are a lot of "please do thises" and "please do thats" in these prompts. As far as I know, whether you ask nicely or not has zero impact on what you get back. But I always say "please" anyway. You never know who might be listening. I actually asked Joanna Penn about this — why does she always say "please" in her prompts? Basically, it seems right to her. And I can't argue.

QUICK-REFERENCE GUIDE

This guide summarizes the most frequently needed prompts by publishing stage for easy lookup.

Stage	Go-To Prompts
Getting started	A1.1, A1.3, A1.4
Concept development	A2.1, A2.2, A2.5, A2.6
Outlining	A3.1, A3.2, A3.4, A3.5
Drafting nonfiction	A4.1, A4.2, A4.3, A4.5
Drafting fiction	A5.1, A5.2, A5.3, A5.5
Titles and headings	A6.1, A6.2, A6.4
Front/back matter	A7.1, A7.2, A7.3
Book descriptions	A8.1, A8.2, A8.3
Email marketing	A9.1, A9.2
Social media	A10.1, A10.2

Stage	Go-To Prompts
Media and speaking	A11.2, A11.3
Repurposing	A12.1, A12.3, A12.4
Editing	A13.1, A13.2, A13.6

1: SETTING UP YOUR AI SESSIONS

These prompts establish context so the AI understands who you are and what you need before you dive into the work. Use them at the beginning of a new session or whenever you're switching from one project to another.

Enter project-specific information everywhere you see brackets and uppercase labels like YOURNAME or BRIEF DESCRIPTION OF BOOK.

A1.1 — Establishing Your Author Identity

You are helping a self-published author. Here is the context you need:
My name is [YOUR NAME]. I write [GENRE/SUBJECT AREA] books for [DESCRIBE YOUR TARGET READER – e.g., "women in their 40s navigating career change" or "hobbyist woodworkers with intermediate skill"].
My writing style is [DESCRIBE YOUR TONE – e.g., "warm and conversational," "authoritative but accessible," "dry and witty"].
My current project is [BRIEF DESCRIPTION OF BOOK]. Please keep this context in mind as we work together. Confirm that you understand before we begin.

A1.2 — Introducing a Manuscript in Progress

I'm going to share the first [NUMBER] pages/chapters of my manuscript with you. I'd like you to read it carefully and then be prepared to help me continue writing in the same voice, tone, and style.
Please do not summarize or comment on it yet – just confirm you've absorbed it and are ready to assist. [PASTE TEXT]

A1.3 — Setting a Consistent Voice Standard

Here is a sample of my writing that represents the voice I want to maintain throughout this book. [PASTE 2–3 PARAGRAPHS].
For everything we work on together, please match this voice as closely as possible — the sentence rhythm, the vocabulary level, the degree of formality, and the way I address the reader. If I drift from it, feel free to point that out.

A1.4 — Session Reset Prompt

Before we start today's session, here's a quick recap of my project:
I'm writing a [GENRE] book titled [WORKING TITLE], aimed at [TARGET READER]. So far I've completed [WHAT'S DONE].
Today I want to focus on [TODAY'S SPECIFIC GOAL]. Please hold this context as we work.

2: DEVELOPING YOUR BOOK CONCEPT

Use these prompts before you've committed to a book idea, or when you're trying to sharpen a concept you already have.

A2.1 — Pressure-Testing a Book Idea

I have an idea for a nonfiction/fiction book. Here it is: [DESCRIBE YOUR IDEA IN 2–4 SENTENCES].
Please evaluate this idea from the perspective of a seasoned book editor. Tell me: What's genuinely strong about it?
What's unclear or underdeveloped?
Who is the most likely reader?
What competing books already exist in this space, and how might mine stand apart?
Be honest – I need useful feedback, not encouragement.

A2.2 — Identifying Your Target Reader

I'm writing a book about [TOPIC]. Help me define my ideal reader in detail.
Who are they?
What is their knowledge level on this subject?
What problem are they trying to solve, or

what experience are they seeking?
What do they already believe about this topic, and where might they be skeptical?
What other books do they read?
What would make them choose my book over others?

A2.3 — Generating Competing Book Concepts

I want to explore different angles for a book about [BROAD TOPIC].
Please generate five distinct book concepts within this subject area — each with a different focus, a different target reader, and a different structural approach.
For each, suggest a working title and a one-sentence description of who it's for and what it delivers.

A2.4 — Narrowing a Too-Broad Topic

My book topic is [BROAD TOPIC], but I suspect it's too broad to be useful or marketable as a single book.
Help me identify five more specific, focused subtopics within this area that could each stand alone as a compelling, well-defined book.
For each, explain what makes it distinct and who would most want to read it.

A2.5 — Finding Your Book's Core Promise

I'm writing a book about [TOPIC] for [TARGET READER].
Help me articulate the core promise of this book — the single most important thing a reader will gain by reading it.
This should be specific, meaningful, and something a reader could not easily get elsewhere.
Give me five different versions of this promise, ranging from practical to emotional in tone.

A2.6 — Validating a Fiction Premise

I have an idea for a [GENRE] novel. Here is the premise: [DESCRIBE YOUR PREMISE]. Please evaluate it from the perspective of an experienced fiction editor.
What's compelling about it?
What questions does it raise that the story will need to answer?
What genre conventions does it align with, and where does it break from them?
What are the potential weaknesses, and how might they be addressed?

3: BUILDING YOUR OUTLINE

A3.1 — Full Nonfiction Outline

Help me build a detailed chapter-by-chapter outline for a nonfiction book with the following parameters:

Title (working): [TITLE]
Subject: [TOPIC]
Target reader: [READER DESCRIPTION]
Core promise: [WHAT THE READER WILL GAIN]
Approximate length: [NUMBER] chapters
Tone: [TONE DESCRIPTION]
For each chapter, provide a title, a 2–3 sentence description of the content, and the key takeaway for the reader. The chapters should build logically on one another and create a sense of journey from beginning to end.

A3.2 — Full Fiction Outline (Three-Act Structure)

Help me build a scene-by-scene outline for a [GENRE] novel using a three-act structure. Here are the parameters:

Working title: [TITLE]
Premise: [PREMISE]
Main character: [NAME, brief description]
Central conflict: [DESCRIBE]
Setting: [TIME AND PLACE]
Approximate word count target: [NUMBER]

For each act, provide a breakdown of major scenes or chapters, the key turning points, and how the character changes from beginning to end.

A3.3 — Alternative Outline Structures for Nonfiction

I'm writing a nonfiction book about [TOPIC] for [READER].
I'd like to see three different ways to structure this book – three distinct organizational approaches that would each work for this content.
For each approach, explain the logic behind it and what kind of reader it would suit best.
My current thinking is [DESCRIBE YOUR CURRENT APPROACH], but I want to make sure I'm not missing a better option.

A3.4 — Filling Gaps in an Existing Outline

Here is my current outline for my book: [PASTE OUTLINE].
Please review it and identify: any gaps in the content where something important seems to be missing; any chapters that feel redundant or could be combined; any logical sequence problems where a chapter might work better in a different position; and any topics a reader would expect to find that aren't addressed.
Please be direct.

A3.5 — Chapter-Level Breakdown

I need to develop Chapter [NUMBER] of my book more fully. The chapter is titled "[CHAPTER TITLE]" and its purpose is to [DESCRIBE WHAT THIS CHAPTER ACCOMPLISHES].
Please break it down into sections and subsections, suggest what content belongs in each, and recommend an approximate word count for each section.
The chapter should run approximately [NUMBER] words total.

4: DRAFTING NONFICTION CONTENT

A4.1 — Drafting a Chapter Section

Please write a [APPROXIMATE LENGTH — e.g., "400-word"] section for my nonfiction book. Here is the context:

Book topic: [TOPIC]
Target reader: [READER]
Tone: [TONE]
This section covers: [DESCRIBE WHAT THIS SECTION SHOULD EXPLAIN OR ACCOMPLISH]
Key points to include: [LIST 3–5 POINTS]
Do not include: [ANYTHING TO AVOID]
Write in a style that is [TONE DESCRIPTION] and assumes the reader [KNOWLEDGE LEVEL].

A4.2 — Writing an Opening Hook

Write five different opening paragraphs for a chapter titled "[CHAPTER TITLE]" in a book about [TOPIC] for [TARGET READER]. Each opening should use a different technique: (1) a surprising statistic or fact, (2) a relatable scenario the reader has probably experienced, (3) a thought-

provoking question, (4) a brief anecdote, and (5) a bold declarative statement. I'll choose the one that fits my voice best.

A4.3 — Generating Examples and Analogies

I'm writing about [CONCEPT] for readers who are [DESCRIBE READER'S BACKGROUND]. This concept can be difficult to grasp.
Please give me: three real-world examples that illustrate it clearly; two analogies that make it relatable to someone without specialized knowledge; and one brief scenario — a mini case study — that shows the concept in action. Keep the language accessible.

A4.4 — Explaining a Complex Concept Simply

I need to explain [COMPLEX CONCEPT OR TERM] to readers who are [DESCRIBE KNOWLEDGE LEVEL].
Please write a plain-language explanation of [200–300 words] that is accurate but avoids jargon. If technical terms are unavoidable, define them within the text. The tone should be [TONE].

A4.5 — Writing a How-To Section

Please write a step-by-step how-to section for my book that explains how to [SPECIFIC TASK OR PROCESS].
The reader is [DESCRIBE READER].
Assume they are starting from [BEGINNING/INTERMEDIATE] knowledge.
Write [NUMBER] steps, with each step clearly labeled and explained in 2–4 sentences. Include a brief note at the end about the most common mistake readers make when attempting this, and how to avoid it.

A4.6 — Creating Sidebars and Call-Out Boxes

Based on the following section of my book, please create [NUMBER] sidebar or call-out box options.
Each should be self-contained, add value without being essential to the main text, and run approximately [LENGTH] words. Options might include a quick tip, a definition, a related resource, a reader exercise, or a "did you know" fact. [PASTE RELEVANT SECTION]

A4.7 — Writing Reader Exercises and Reflection Questions

At the end of Chapter [NUMBER] in my book about [TOPIC], I want to include [NUMBER] exercises or reflection questions to help readers apply what they've just learned.

The chapter covers [BRIEF SUMMARY].

The reader is [DESCRIBE READER].

Please write exercises that are practical and specific — not vague or generic. Some should be thinking exercises, and at least two should involve a concrete action the reader can take right away.

A4.8 — Drafting a Chapter Summary

Please write a summary for Chapter [NUMBER] of my book. The chapter covers the following key points: [LIST POINTS]. The summary should be approximately [LENGTH] words, be written in [TONE], and reinforce the most important takeaways without simply listing them. It should feel like a satisfying conclusion to the chapter, not a table of contents review.

A4.9 — Building a Glossary

Based on the following content from my book about [TOPIC], please identify all terms that a [TARGET READER] might not know and compile them into a glossary. For each term, write a clear, jargon-free definition of 1–3 sentences. List them alphabetically. [PASTE CONTENT OR OUTLINE]

A4.10 — Fact-Check Prompting

I've written the following passage for my book: [PASTE PASSAGE].
Please review it for factual accuracy. Flag anything that you are uncertain about, anything that may have changed recently, or anything that a reader might reasonably challenge.
Do not simply confirm that it sounds reasonable – be specific about what you know to be accurate, what you are uncertain of, and what I should independently verify.

5: DRAFTING FICTION CONTENT

A5.1 — Creating a Character Profile

Help me develop a detailed character profile for the [PROTAGONIST/ANTAGONIST/SUPPORTING CHARACTER] of my [GENRE] novel.
Here's what I know so far: [BRIEF DESCRIPTION].
Please expand this into a full profile that includes: physical appearance, age, and background; personality traits, strengths, and fatal flaws; core motivation and deepest fear; speech patterns and mannerisms; relationship to the central conflict; and one surprising detail that makes them feel three-dimensional and unexpected.
Write this as a reference document I can use throughout the writing process.

A5.2 — Writing a Scene from Scratch

Please write a scene for my [GENRE] novel with the following parameters:

Characters present: [LIST WITH BRIEF DESCRIPTIONS]

Setting: [DESCRIBE TIME AND PLACE IN DETAIL]
What happens in this scene: [DESCRIBE THE ACTION OR EVENT]
What must be established or advanced: [PLOT POINT, RELATIONSHIP DYNAMIC, THEME, ETC.]
Tone: [TENSE, ATMOSPHERIC, HUMOROUS, ETC.]
Approximate length: [NUMBER] words
Write in close third-person/first-person [CHOOSE ONE] point of view from [CHARACTER NAME]'s perspective.

A5.3 — Writing Dialogue

Write a dialogue scene between [CHARACTER A] and [CHARACTER B] in my [GENRE] novel.
Here's the context: [DESCRIBE THE SITUATION AND WHAT EACH CHARACTER WANTS FROM THIS CONVERSATION].
The scene should feel natural and subtext-rich – neither character says exactly what they mean.
The dialogue should reveal character and advance [DESCRIBE WHAT IT NEEDS TO ADVANCE]. Run approximately [LENGTH].
Avoid dialogue tags other than "said" unless absolutely necessary.

A5.4 — Developing a Setting

Help me develop the setting for a key location in my novel:
[DESCRIBE THE LOCATION]. This place is important because [EXPLAIN ITS ROLE IN THE STORY].
Please write a rich, sensory description of approximately [LENGTH] words that brings this setting to life. Include what the characters see, hear, smell, and feel here.
Also suggest two or three ways this specific setting can actively influence or complicate the events that take place in it.

A5.5 — Writing Past a Plot Problem

I'm stuck in my novel. Here's where I am:
[DESCRIBE THE CURRENT STATE OF THE STORY, THE CHARACTERS INVOLVED, AND WHAT HAS JUST HAPPENED].
The problem is [DESCRIBE WHAT'S NOT WORKING – e.g., "I don't know how my character would believably escape this situation" or "this scene has no tension"].
Please suggest five different ways I might move forward from this point.
Give each option a brief description and explain what it would open up for the rest of the story.

A5.6 — Maintaining Series Continuity

I'm writing Book [NUMBER] in a series. Here is a summary of what has happened in previous books, including established character traits, world-building details, and plot threads that are still open: [PASTE YOUR SERIES BIBLE OR SUMMARY].
I'm now working on [DESCRIBE CURRENT SCENE OR CHAPTER].
Please help me ensure this new content is consistent with what's been established. Flag any contradictions you notice and suggest how to address them.

A5.7 — Pacing and Tension Check

Please read the following scene from my novel and evaluate its pacing and tension: [PASTE SCENE]. Tell me:
Does it move too fast, too slow, or unevenly?
Where does tension build effectively, and where does it sag?
Are there places where the prose slows the action unnecessarily?
Suggest specific revisions that would improve the energy of this scene without changing its core events.

A5.8 — Genre Conventions Check

I'm writing a [SPECIFIC GENRE] novel. Based on established conventions of this genre, what are the key elements readers expect?
What tropes are considered essential, and which ones are considered overused?
Where does my premise – [DESCRIBE PREMISE] – align with or deviate from these conventions?

Help me understand where I have flexibility and where deviating from convention might put readers off.

6: TITLES, SUBTITLES, AND HEADINGS

A6.1 — Generating Title Options

Please generate twenty title options for a [GENRE/TYPE] book about [TOPIC] for [TARGET READER].
The book's core promise is [CORE PROMISE].
Mix approaches: some should be punchy and short, some should be intriguing or evocative, some should be clearly descriptive, and a few should take a risk. Do not play it safe — I'd rather have interesting options to react to than a list of safe ones.

A6.2 — Developing a Subtitle for Nonfiction

The working title for my nonfiction book is "[TITLE]."
Please generate fifteen subtitle options.
The book is about [TOPIC], for [TARGET READER], and promises [WHAT READER WILL GAIN].
A good subtitle for this book should clarify the topic, identify the reader, suggest the

outcome, and be searchable online.
Show me a range from highly practical to slightly more intriguing.

A6.3 — Testing a Title's Strength

I'm considering this title for my book: "[TITLE]."
Please evaluate it honestly. Is it clear?
Is it memorable?
Does it suggest the right audience?
Does it have strong search potential?
Could it be misread or misunderstood?
How does it compare, in general terms, to successful titles in [GENRE/SUBJECT AREA]?
What are its biggest weaknesses, and how might I address them?

A6.4 — Writing Chapter Titles

Please write compelling chapter titles for the following chapters in my book. For each, I'll give you the chapter's main topic and purpose.
I want titles that are specific enough to be informative but interesting enough to make a reader want to read the chapter.
Avoid generic titles like "Introduction to X" or "Understanding Y." [LIST EACH CHAPTER'S TOPIC AND PURPOSE]

7: FRONT AND BACK MATTER

A7.1 — WRITING A PREFACE

Please write a preface for my book titled "[TITLE]."
The preface should be approximately [LENGTH] words and written in first person. It should tell the story of why I wrote this book, what experience or expertise I bring to the subject, who this book is for, and what a reader can expect to take away. The tone should be [TONE – e.g., "personal and warm," "authoritative but approachable"]. Here are the key points I want to make: [LIST 3–5 POINTS].

A7.2 — WRITING AN INTRODUCTION THAT HOOKS

Write an introduction for my nonfiction book titled "[TITLE]."
The introduction should be approximately [LENGTH] words and accomplish the following:
Hook the reader immediately with a compelling opening, establish the problem or opportunity the book addresses,

introduce the reader to what they will learn and how the book is organized, and make a clear case for why they should read this book now.
Tone: [TONE].
Target reader: [READER DESCRIPTION].

A7.3 — Writing an Author Bio

Please write [NUMBER] versions of my author bio, ranging from 50 to 200 words, for the following uses: a book's back matter, an Amazon Author Page, a speaker introduction, and a social media profile. Here is the information to draw from: [PROVIDE YOUR BACKGROUND, CREDENTIALS, PUBLISHED BOOKS, RELEVANT PERSONAL DETAILS, AND ANY WEBSITES OR SOCIAL HANDLES]. The tone should be [TONE — e.g., "professional but personable"]. Write in third person unless I specify otherwise.

A7.4 — Writing Acknowledgments

Help me write the acknowledgments section for my book.
Here are the people and organizations I want to thank, along with brief notes on why: [LIST PEOPLE AND REASONS].
Please write this in a warm, genuine tone

that sounds like me — not like a corporate press release. Aim for approximately [LENGTH] words.
My overall tone throughout the book is [TONE], so please match that.

A7.5 — Creating a Resources and Recommended Reading List

Based on the following topics covered in my book, please suggest categories of resources — books, websites, organizations, and tools — that I should consider including in a Resources section. [LIST YOUR MAIN TOPICS].
For each category, describe what kind of resource would be most useful. I will then research and verify specific titles and links myself.

8: BOOK DESCRIPTIONS AND MARKETING COPY

A8.1 — Writing a Back Cover Description (Nonfiction)

Please write three versions of a back cover book description for my nonfiction book. Here are some details:

Title: [TITLE]
Topic: [TOPIC]
Target reader: [READER]
Core promise: [WHAT THE READER GAINS]
Key topics covered: [LIST 4–6]
My credentials: [BRIEF AUTHOR CREDENTIALS]
Each version should be approximately 150–200 words. Version 1 should lead with the reader's problem. Version 2 should lead with the book's promise. Version 3 should lead with a surprising hook or statement. All three should end with a compelling reason to buy now.

A8.2 — Writing a Back Cover Description (Fiction)

Please write three versions of a back cover description for my [GENRE] novel. Here are the details:
Title: [TITLE]
Premise: [BRIEF PREMISE]
Main character: [NAME AND BRIEF DESCRIPTION]
Central conflict: [DESCRIBE]
Tone/feel: [E.G., "dark and atmospheric," "fast-paced and funny"]
Do NOT reveal: [WHAT TO KEEP HIDDEN]
Each version should be 100–150 words. They should intrigue without over-explaining, create urgency without feeling cheap, and end with a question or statement that makes not reading the book feel like a missed experience.

A8.3 — Amazon Product Description with Keywords

Please write an Amazon product description for my book titled "[TITLE]." This is a [GENRE/TYPE] book for [TARGET READER].
The description should be 300–500 words, use natural language that incorporates these search keywords: [LIST YOUR

TARGET KEYWORDS], include a bulleted list of 5–7 reader benefits, and end with a clear call to action.
Format it using Amazon's preferred style: short paragraphs, bold opening line, benefits-focused language.

A8.4 — Series Description

I publish a series of books titled "[SERIES NAME]."
The series is designed for [TARGET READER] and covers [OVERALL TOPIC AREA]. Individual books in the series include: [LIST TITLES AND ONE-SENTENCE DESCRIPTIONS].
Please write a series description of approximately 150 words that I can use across all marketing materials to help readers understand what the series offers as a whole and why they should explore all the titles.
Tone: [TONE].

A8.5 — Press Release

Please write a press release announcing the publication of my new book, "[TITLE]." Here are the details:

Author name: [NAME]
Publication date: [DATE]
Publisher: [PUBLISHER NAME]
Price and formats available: [DETAILS]
ISBN: [NUMBER]
Where to buy: [LINKS OR RETAILERS]
Book description (one paragraph): [PASTE OR DESCRIBE]
Author bio (two sentences): [PASTE]
Contact information: [DETAILS]
Write in standard press release format with a strong headline, dateline, and boilerplate.
The tone should be professional.
Include one brief fictional quote from the author that sounds authentic and not promotional.

A8.6 — Advance Reader Copy (ARC) Request Letter

Please write a letter I can send to potential advance readers requesting that they read and review my upcoming book, "[TITLE]." The book is about [BRIEF DESCRIPTION]

and publishes on [DATE]. I'm looking for reviewers who are [DESCRIBE IDEAL REVIEWERS].
The letter should be warm but professional, briefly describe the book, explain what I'm asking of them (read the ARC and post an honest review), give them a deadline, tell them how they'll receive the ARC, and thank them genuinely. Length: approximately 200–250 words.

9: EMAIL AND NEWSLETTER MARKETING

A9.1 — Launch Announcement Email

Write a launch announcement email to my reader list announcing my new book, "[TITLE]."
My list knows me as [DESCRIBE YOUR RELATIONSHIP WITH YOUR LIST – e.g., "a writing coach who sends weekly tips"].
The book is about [BRIEF DESCRIPTION].
The tone should be [TONE – e.g., "excited but not breathless, personal not salesy"].
Include: a personal opening that doesn't lead with the book, a natural transition into the announcement, a description of who the book is for and what it does, a clear purchase link placeholder, and a P.S. that creates a small sense of urgency without being pushy. Length: 250–350 words.

A9.2 — Email Nurture Sequence (5-Part)

I've just launched a new book titled "[TITLE]" about [TOPIC].
Please write a five-email nurture sequence for subscribers who signed up for my email list but have not yet purchased the book.
Each email should have a different focus:
Email 1 (Day 1): Welcome and deliver on the lead magnet promise
Email 2 (Day 3): Share a useful tip or insight related to the book's topic

Email 3 (Day 5): Tell a relevant personal story or case study
Email 4 (Day 7): Address the most common objection to buying the book
Email 5 (Day 10): Soft pitch with testimonial or social proof
My tone is [TONE]. Target reader: [READER DESCRIPTION]. Keep each email to 200–300 words.

A9.3 — Newsletter Content Around a Book Launch

I publish a regular newsletter on the topic of [TOPIC].
I have a new book launching in [TIMEFRAME].
Please help me plan four newsletter issues that build excitement for the launch without feeling like pure promotion.
For each issue, suggest: a subject line, a primary topic that serves my readers, and a natural way to connect it to the upcoming book without the whole issue feeling like an ad.

10: SOCIAL MEDIA CONTENT

A10.1 — Social Media Launch Posts (Multi-Platform)

Please write social media posts announcing the launch of my new book, "[TITLE]," for the following platforms.

Adapt the tone and length for each:
Facebook (personal, conversational, 100–150 words)
Instagram (visual-forward, use line breaks, 80–100 words plus 10 relevant hashtags)
LinkedIn (professional, credibility-focused, 150–200 words)
X/Twitter (punchy, under 280 characters, two versions)
The book is about [BRIEF DESCRIPTION], for [TARGET READER].
My tone is generally [TONE].

A10.2 — Content Calendar for a Book Launch Month

Please create a 30-day social media content calendar for the launch of my book, "[TITLE]."
The book is about [TOPIC] for [TARGET

READER]. It launches on [DATE].
I post primarily on [PLATFORMS].
The calendar should include a variety of content types — tips, quotes, behind-the-scenes, reader questions, testimonials, direct promotion, and evergreen content — with no more than 30% of posts being direct sales pitches.
Give me a brief description of each post, not the full caption.

A10.3 — Repurposing Book Content for Social Media

I want to repurpose content from my book, "[TITLE]," into social media posts.
Here is an excerpt: [PASTE 300–500 WORDS].
Please create ten social media posts from this content — a mix of short insights, quotable statements, questions that invite engagement, and tips.
Format each for a general social media audience, and flag which two or three you think would perform best and why.

11: SPEAKING, PODCASTS, AND MEDIA

A11.1 — Speaker One-Sheet

Please help me write a speaker one-sheet for appearances related to my book, "[TITLE]."
Include: a compelling speaker bio of approximately 150 words, three to five talk topics derived from the book's content [DESCRIBE MAIN THEMES], a suggested audience description, and three testimonial placeholders.
Tone: professional but personable.
The information to draw from: [PASTE YOUR BIO AND BOOK DESCRIPTION].

A11.2 — Podcast Pitch Letter

Please write a pitch letter I can send to podcast hosts, proposing me as a guest. The letter should be concise (under 250 words), reference the podcast by name and show I've done my research [I WILL PERSONALIZE THIS], explain who I am and what I've written, propose three specific episode topics that would serve their audience, and include a brief, confident closing. My book is "[TITLE]" about [TOPIC]. My background: [BRIEF CREDENTIALS].

A11.3 — Podcast Interview Prep

I'm going on a podcast to discuss my book, "[TITLE]."
The show focuses on [PODCAST TOPIC/AUDIENCE].
Please generate: ten questions the host is likely to ask me, three questions I should hope they ask (because my answers would be compelling), two questions I should be prepared for that might be challenging, and my ideal "closing statement" when they ask where people can find me and my work.
My book is about [TOPIC] and my target reader is [READER DESCRIPTION].

A11.4 — Conference Talk Abstract

Please write an abstract for a conference talk based on my book, "[TITLE]."
The conference focuses on [CONFERENCE TOPIC], and the audience is [AUDIENCE DESCRIPTION].
The talk should run [LENGTH] minutes.
The abstract should be 150–200 words, include a compelling title for the talk, describe what attendees will learn, and make clear why this topic matters to this specific audience.
Please also write a 50-word version for the conference program.

12: REPURPOSING YOUR BOOK

A12.1 — Online Course Outline

My book, "[TITLE]," covers [TOPIC] for [TARGET READER].
I want to repurpose it as an online course. Please help me structure a [NUMBER]-module course based on the book's content.
For each module, suggest: a title, the key lesson or outcome, two to three lessons or video topics within the module, and one exercise or assignment.
The course should feel like a transformation — not just the book read aloud.

A12.2 — Workbook Development

I want to create a companion workbook for my book, "[TITLE]."
The book covers [TOPIC] and has [NUMBER] chapters. For each chapter, please suggest: two to three reflection questions, one practical exercise, and one "commitment" prompt where the reader commits to a specific action.

The workbook should help readers move from reading to doing.

A12.3 — Blog Post Series from Book Content

My book, "[TITLE]," covers [TOPIC].
Please suggest a 12-post blog series that draws from the book's content without giving the whole book away. For each post, give me a title, a one-paragraph description of the content, and a note on how it connects to the book without requiring the reader to have read it.
These posts should work both as standalone pieces and as a series.

A12.4 — Lead Magnet Ideas

I'm looking for lead magnet ideas – free downloadable resources I can offer to grow my email list – based on my book, "[TITLE]," about [TOPIC].
Please suggest eight different lead magnet concepts that would attract my ideal reader, [READER DESCRIPTION].
For each, give a title, a brief description of what it contains, and why it would appeal to this reader.
Include a range of formats: checklists,

guides, templates, quizzes, mini e-books, and so on.

13: EDITING AND REFINEMENT

A13.1 — DEVELOPMENTAL EDIT REQUEST

Please read the following chapter from my manuscript and evaluate it at a developmental level – not grammar or spelling, but structure, clarity, and impact. [PASTE CHAPTER].
Tell me: Does the chapter accomplish what it sets out to do? Is the structure logical? Are there gaps, digressions, or redundancies? Does it begin and end effectively? What would an editor likely flag as the most important issues to address in a revision?

A13.2 — LINE EDITING FOR CLARITY

Please line-edit the following passage for clarity, concision, and flow.
My goal is not to change the meaning or my voice, but to remove unnecessary words, untangle confusing sentences, and make the prose easier to read.
Track your changes by showing me the original and revised version of each paragraph and briefly explain your

reasoning for significant changes. [PASTE PASSAGE]

A13.3 — Voice Consistency Check

I'm concerned that my writing voice has shifted across different parts of my manuscript.
Please read the following two sections – one from early in the book and one from later – and tell me whether the voice, tone, vocabulary level, and sentence rhythm are consistent.
If they differ, explain specifically how and suggest which direction I should standardize toward. [PASTE BOTH SECTIONS WITH LABELS]

A13.4 — Strengthening a Weak Opening

The opening of my chapter/book isn't working.
Here is what I currently have: [PASTE CURRENT OPENING].
Here is what this chapter/book is about: [BRIEF DESCRIPTION].
Please rewrite this opening three ways – each using a different technique – that would more effectively hook the reader.
After presenting the options, explain why

each might work and what the tradeoff of each approach is.

A13.5 — Tightening Overwritten Prose

The following passage is overwritten — too many words, too many adjectives, and the meaning is getting buried.
Please rewrite it so it's tighter and more direct without losing its essential meaning or my voice.
Aim to reduce the word count by at least 25%. [PASTE PASSAGE]

A13.6 — Final Proofread Check

Please proofread the following text carefully for spelling errors, grammatical mistakes, punctuation problems, inconsistent capitalization, and any repeated words or phrases that appear too close together.
List each issue you find with the original text and your suggested correction. Do not suggest stylistic changes — only flag clear errors. [PASTE TEXT]

GLOSSARY: AI TERMS EVERY AUTHOR SHOULD KNOW

Like everything else related to AI, these terms may change in meaning and importance over time.

Term	In Plain English
AI / Artificial Intelligence	Technology that processes and generates language
Large Language Model (LLM)	The engine powering tools like ChatGPT and Claude
Prompt	The instruction you give the AI
Context Window	How much text the AI can hold in memory at once
Hallucination	When AI invents facts and states them confidently

Term	**In Plain English**
Verification	Checking AI's facts before you publish them
Iteration	Refining AI output through conversation
Training Data	The text the AI learned from
Knowledge Cutoff	The date beyond which the AI has no information
Temperature	How creative vs. predictable the AI's output is
Token	The unit of text AI processes (≈ ¾ of a word)
Model Version	Which generation of the AI you're using
Memory	Whether the AI remembers previous sessions
Content Filter	Why the AI sometimes declines a request

Term	**In Plain English**
Output	Everything the AI generates in response
Input	Everything you provide to the AI
Bias	Systematic preferences baked into AI output
Fine-Tuning	Training a model for a specific purpose
Generative AI	AI that creates new content, not just analyzes it
Voice	Your distinctive writing style — yours to protect

ABOUT THIS BOOK

Say "Hello" to AI in action.

I began fooling around with AI more than five years ago. Non-plussed by the results, I waited a few months to try again. I kept at my explorations until early in 2025, when an opportunity arose that cried out for an AI solution. AI had gotten much, much better.

Who Better to Write a Book on AI?

When we decided to add a book about AI-assisted writing and publishing to our *Publish It Yourself!* Series, the obvious (to me) question was, ***"Who better to help draft this book than AI?*** Indeed.

The result was Book 3 in the *Publish It Yourself!* Series, *Artificially Intelligent: AI for Self-Publishers.*

We asked our new BFF Claude to flesh out a detailed outline based on a very detailed prompt we painstakingly wrote. We followed this up by asking for help with developing the sections, chapters, and appendices. The outputs were right on the money, mostly, and amazing.

We edited what we got from Claude extensively and added our own stories and flavor. We omitted pages of material out of scope for this particular title.

This book exists to help you amp up your writing and publishing with assistance from AI and get you where you want to go faster and more productively. If we've tried to make one thing clear in this book, it's that better prompts make for better, more discrete and precise results. Let us know about your experience at mmc@touchwoodpress.com.

ABOUT THE AUTHOR

Michael McCandless has worked in and around book publishing for more than 30 years across traditional publishing, textbook, technical, nonprofit, and independent self-publishing markets.

Mike's pioneering work across several periods of rapid technological change laid the groundwork for the Touchwood Press business.

He founded Touchwood Press in 2013 as an independent publishing imprint focused on practical, ethical, and sustainable approaches to self-publishing. He teaches self-publishing workshops for first-time authors, helping them understand not just how to publish a book, but what publishing involves as a tradition, industry, and system.

Mike's approach to writing and publishing emphasizes clarity over hype, realistic expectations, respecting the creative process, and encouraging nervous self-publishers.

Reach out to Mike at
mmc@touchwoodpress.com.

ABOUT THE PUBLISH IT YOURSELF SERIES

The *Publish-It-Yourself* series is designed to help writers move from manuscript to finished book with clarity, confidence, and realistic expectations.

Rather than offering shortcuts or universal formulas, each volume focuses on a distinct stage of the independent publishing journey. Together, the books provide context, instruction, and practical guidance—while allowing readers to engage at the level that fits their experience and goals.

How the Series Is Structured

Each book in this series serves a specific purpose:

Book 1—***Ready, Set, Publish!: Basic Training for Self-Publishers***. An orientation to modern publishing, this volume explains how publishing works, what decisions authors face, and why those decisions matter—before tools or platforms enter the picture.

Book 2—***Publish-It-Yourself!: Step-by-Step to Your Self-Published Book***. A guide through the tasks involved in self-publishing your books using the Amazon Kindle Direct Publishing system. This volume assumes familiarity with the broader publishing concepts introduced in the Primer.

Book 3—***Artificially Intelligent: AI for Self-Publishers***. All aboard the AI Express—the big picture.

Book 4—***AI Prompts for Self-Publishers: Collaborating with Your New Publishing Assistant***. From concept development to launch and promotion, AI can help with the right prompting.

The *Publish-It-Yourself* Course—Available Online or Live

Coaching—By arrangement.

MORE BOOKS FOR WRITERS

Writing Starters: Appetizers for Your Brain. Two Volumes. These books each provide 52 ideas to get you writing. From *Write a Letter to a Favorite* Teacher to *Are You Creative?*, the idea is that only actual writing is writing. Use *Writing Starters* to help generate new ideas or just to fine-tune your style and voice. Use the books when you can't seem to finish your project or anytime to "sharpen the saw."

In Development

Book 5—***Book Formatting for Self-Publishers: Building Your Book from Scratch***

Book 6—***It's All Marketing: Finding and Serving Your Readers***

Book 7—***There's No Business Like Publishing: Growing Your Self-Publishing Enterprise***

Each book is written to stand on its own while contributing to a coherent whole. The *Publish-It-Yourself* Series benefits from our experience with traditional and digital publishing and working with first-time authors as an instructor, coach, and service provider.

Get your writing out. Tell your story. Fulfill your creative dreams. Publish it yourself!

Learn more at publish-it-yourself.com.

ABOUT THE PUBLISHER

Touchwood Press is an independent, or "indie," publisher of fiction and nonfiction books and a service provider helping writers, businesses, and organizations to publish their intellectual property. The *Publish-It-Yourself!* series includes books, live and online courses, and coaching services designed to provide a personalized level of guidance and support to new and established authors.

Our clients include first-time writer/authors and businesses and organizations tasked with transforming intellectual property such as training materials or marketing collateral into digital media for sale or distribution in support of brand development.

Contact publisher@touchwoodpress.com to learn more about how we can help with your project. Learn more at touchwoodpress.com.

Made in the USA
Coppell, TX
06 November 2026

11491037R00041